Photographic reproduction of the original covers, completely unmarked save for the stamped numeral "2".
NAA: A3269, Q3

THE PARTISAN LEADER'S HANDBOOK

G.S.(R), THE WAR OFFICE

MAY, 1939

ISBN-13: 978-1976335921
ISBN-10: 1976335922

(©) C.A. Brown, 2017
All Rights Reserved

G.S.(R)

CONTENTS

Introductory vii

Principles of Guerilla Warfare and Sabotage 1

Road Ambush Appendix I 11

Rail Ambush Appendix II 17

The Destruction of an Enemy Post,
Detachment or Guard Appendix III 25

Concealment and Care of Arms and
Explosives Appendix IV 29

The Enemy's Information System
and how to Counter It Appendix V 31

How to Counter Enemy Action Appendix VI 33

Guerilla Information Service Appendix VII 35

Sabotage Methods Appendix VIII 37

G.S.(R)

Introductory.

General Service (Research), or *G.S.(R)*, was the rather bland designation for what would later evolve into the "dirty tricks" department of the British War Office. Established in 1936, G.S.(R) was, by January 1939, tasked with development of various special operations, military intelligence, special weapons and equipment, and irregular warfare concepts in the years immediately preceding the outbreak of the Second World War.

Redesignated *Military Intelligence (Research)*, or *M.I.(R)*, in the Spring of 1939, with the crisis in Europe rapidly coming to a head, the organisation's work took on far greater urgency. M.I.(R) was expanded and became responsible for not only covert intelligence and irregular warfare, but also the development of novel anti-tank weapons and tactics in an attempt to blunt the sharp edge of Hitler's state-of-the-art armoured forces which, UK War Office planners had correctly determined, could very well ride roughshod across Europe if left unchecked.

Joining G.S.(R) in April 1939 was Major Colin Gubbins. Gubbins was an artillery officer who had seen service in the Great War in France, and significantly, had been involved in operations against guerrilla forces during the Russian Civil War in 1918/1919 and in Ireland against the IRA from 1919 to 1921. Later, while posted to the Northwestern Frontier in India, he had become familiar with irregular tribal warfare the use of human intelligence assets. He was perfectly placed to author three handbooks for irregular warfare - *The Art of Guerilla Warfare*, *The Partisan Leader's Handbook* and *How to Use High Explosives*.

Gubbins had the three handbooks completed by May of 1939, and then was dispatched to Poland to investigate the establishment of a "stay-behind" guerrilla force in the likely event of a German invasion of that country. As it stood, the lightning speed of the German invasion overtook the planning of the Polish "stay behind" force and Gubbins returned to the UK.

With the outbreak of war, Gubbins was tasked with establishing the UK's Independent Companies (later designated "Commandos"), which he joined in operations in Norway. Later still, Gubbins was tasked with the establishment of a "Stay-Behind" guerrilla force in the UK in advance of a threatened Nazi invasion of Britain.

This force was given the rather nondescript designation of "Auxiliaries" and were nominally placed on the war establishment of the Local Defence Volunteers (Home Guard). The Auxiliaries were formed into local bands called Patrols and were trained in commando tactics and supplied with a variety of weapons and demolitions stores which were cached around the countryside. Operating from underground "operational base" dugouts, patrols would harry and hinder occupying German forces while gathering intelligence which would be transmitted to the British government in exile in Canada via specially trained covert wireless units.

Gubbins was posted to the newly formed Special Operations Executive toward the end of 1940. G.S.(R) had morphed into M.I.(R) and in 1940 was merged with Section D of the Secret Intelligence Service to become the Special Operations Executive (SOE). An unusual organisation, SOE combined intelligence functions and special warfare, making it a self-contained covert force which later inspired the creation of the US Office of Strategic Services (OSS), which later itself inspired the creation of the United States Central Intelligence Agency (CIA). In 1943, Gubbins was appointed head of the SOE and oversaw operations all over the world, from Europe to North Africa to South East Asia and the China/Burma/India theatre.

The Partisan Leader's Handbook was a short training pamphlet which, unlike its larger-in-scope companion volume *The Art of Guerilla Warfare*, is about the "nuts-and-bolts" of unconventional warfare. It was designed as a reference for would-be leaders of guerrilla bands in occupied territory, providing guidance on the conduct of raids, sabotage, demolitions, security, intelligence.

It was influenced not only by Gubbins' own experiences in Russia, Ireland and India, but also by his intensive study of guerrilla warfare in general, from the Boer commandos in South Africa during the Boer War, to the guerrilla operations in the Spanish Civil War and the ongoing Chinese Civil War and Sino-Japanese war, where Communist leader Mao Tse Tung was using guerrilla tactics to fight both the Chinese Nationalists and the invading Japanese.

The pamphlet was later used as the basis for the SOE's irregular warfare training syllabus. *The Partisan Leader's Handbook* and its companion volume, *Art of Guerilla Warfare* are unique in that they represent the first official British military doctrine on the conduct of guerrilla warfare against an occupying force. This they do in a brief and to the point manner, and if one reads closely, it becomes clear that there is

G.S.(R)

still much wisdom contained within which could be of use to insurgent and counterinsurgent alike in the modern era.

CA Brown
August 2017

G. S. (R)

PARTISAN LEADER'S HANDBOOK.

Principles of Guerilla Warfare and Sabotage.

1. Remember that your object is to embarrass the enemy in every possible way so as to make it more difficult for his armies to fight on the main fronts. You can do this by damaging his rail and road communications, his telegraph and postal system, by destroying small parties of the enemy and in many other ways which will be explained later.

   Remember that everything you can do in this way is helping to win freedom again for your people.

2. You must learn the principles of this type of warfare, which are as follows:-

   (a) Surprise is the most important thing in everything you undertake. You must take every precaution that the enemy does not know your plans.

   (b) NEVER ENGAGE IN ANY OPERATION UNLESS YOU THINK SUCCESS IS CERTAIN. BREAK OFF THE ACTION AS SOON AS IT BECOMES TOO RISKY TO CONTINUE.

   (c) Every operation must be planned with the greatest care. A safe line of retreat is essential.

   (d) Movement and action should, wherever possible, be confined to the hours of darkness.

   (e) Mobility is of great importance ; act therefore, where your knowledge of the country and your means of movement -- i.e. bicycles, horses, etc. -- give you an advantage over the enemy.

   (f) Never get involved in a pitched battle unless you are in overwhelming strength.

   (g) Never carry incriminating documents on your person nor leave them where they can be found.

The whole object of this type of warfare is to strike the enemy, and disappear completely leaving no trace; and then to strike somewhere else and vanish again. By these means the enemy will never know where the next blow is coming, and will be forced to disperse his forces to try and guard all his vulnerable points. This will provide you with further opportunities for destroying these small detachments.

3. Types of Operations :— Operations can be divided into two main types:—
- (a) Those of a military nature which entail the co-ordinated action of a certain number of men under a nominated leader.

- (b) Individual acts of sabotage, of sniping sentries etc., for which men can be specially selected to work individually in certain areas.

For action of a military nature the choice of suitable leaders is of great importance, A leader must have courage and resource, he must be intelligent and a good administrator and be a man of quick decision. He must know intimately the country in which he is operating, and should be able to use a compass and map. The sort of man required is the type whom other men will willingly accept to lead them in dangerous actions, and whose personality will hold them together.

The size and composition of guerilla parties must depend on the nature of the country and the hold which the enemy has over it. It must be remembered that the speed of modern communications, i.e., motors, wireless, etc., and the presence of aeroplanes make it very difficult for a large party to remain concealed for any length of time. Parties should therefore number between 8 and 25, depending upon the work to be done; such parties can move quickly and yet hide themselves fairly easily. Under specially favourable conditions, it may be possible to collect several parties together, up to 100 men or more, tor some important undertaking. In such cases, however, the arrangements for dispersal after the operation must be made with special care.

4. Modern large-sized armies are completely dependent on roads, railways, signal communications etc., to keep themselves supplied with food, munitions and petrol, without which they cannot operate. These communications therefore form a most suitable target for guerilla warfare of all kinds, and any attack on them will at once force the enemy to disperse his forces in order to guard them. Communications are open to attacks both of the military and sabotage type. Attacks can also be directed against small detachments of the enemy, stocks of food, munitions, etc., and many other objects.

G.S.(R)

5. Military action is employed when it appears that damage can only be inflicted if force has to be used first. The following are types of military action:-

    (a) Destruction of vital points on roads, bridges, railways, canals, etc., when action by an individual employing secret means would not be effective. If a hostile guard has first to be overpowered, or work preliminary to destruction requires a considerable number of men, the project must be undertaken as a military operation.

    (b) The raiding and destruction of hostile mails, either in lorries or trains.

    (c) The destruction of enemy detachments and guards.

    (d) The organization of ambuscades of hostile troops and convoys travelling by road or train.

    (e) The destruction of stocks and dumps of food, petrol, munitions, lorries, etc., by first overpowering the guards on them.

    (f) The seizure of cash from hostile pay-offices etc.

ETC.   ETC.

6. Military action is greatly facilitated by the support of the local population. By this means, warning can be obtained of all hostile moves, and it will not be possible for the enemy to carry out surprise action. It is therefore important to endeavour not to offend the people of each district, but to encourage their patriotism and hatred of the enemy. Successful action against the enemy will breed audacity and force the people to take note and respond. Their response in the first instance should be directed to the supply of information about the enemy, his strength, movements, etc., and to assistance in the concealment of compatriots who are taking part in guerilla warfare. In effect, the people must be taught to boycott the hostile troops completely, except as may be necessary to obtain information. This can best be done by convincing them that the enemy's occupation is only temporary, that he will soon be ejected, that those of the people who have helped will then be rewarded, but that those who have fraternized with the enemy will be ruthlessly punished. The question of "informers" and traitors who are in league with the enemy is dealt with later.

7. The areas most suitable for military action are those where cover, such as rocks, trees, undergrowth, etc., give a concealed approach to the object or detachment to be attacked. Such cover not only provides an opportunity for

attack without discovery, but also for getting away safely when the attack is completed. In all such attacks, it is important that sentries should be posted on all approaches to give warning of any possible surprise by the enemy; it is not necessary that all these sentries should be armed men, in fact it will frequently be of advantage to use some women and children, who are less likely to be suspected. A simple code of signals must be arranged.

Every operation of this nature must be most carefully planned. When some particular operation has been decided upon, the locality must be thoroughly reconnoitred, and the enemy's movements in the vicinity should be systematically studied and noted over a period of days, with special reference to such points as the following, where applicable:-

(a) Hours when sentries are relieved, and how relief is carried out.

(b) Total strength of guard or detachment.

(c) How and when do supplies for the guard arrive? Are civilians allowed to enter the post?

(d) Where do men not on sentry duty keep their rifles? Are these rifles chained up or in plain racks?

(e) Are men allowed to leave the position for short periods?

(f) How often are guards inspected, by whom and at what times?

(g) What means of communication for the post exist, i.e., telegraph, motor-cycle, or cycle messenger, carrier pigeons, etc. Can these be destroyed?

(h) Do mails or small detachments of men follow regular routes at fixed times, giving opportunities for ambushing?

(i) Do these detachments have sentries, advance parties, etc., or do they proceed in one group?

(j) Are motor vehicles fitted with bullet-proof or puncture-proof tyres, armoured sides, etc. ?

(k) What special tools and explosives, if any, are required for the operation, and what amount?

Examples of such operations are given at the end of the book.

G . S . (R)

8. Sabotage deals with the acts of individuals or small groups of people, which are carried out by stealth and not in conjunction with armed forces. These undertakings, however, frequently produce very valuable results and, like military action, force the enemy to disperse his strength in order to guard against them. The following are examples of this type of work:—

(a) Jamming of railway points.

(b) Destructive work on roads, railways, canals, telegraphs, etc., where this can be done by stealth.

(c) Firing of stocks of petrol; burning garages, aeroplane hangars, etc

(d) Contamination of food, of forage, etc., by acid, by bacilli, poison, etc.

(e) Contamination of petrol by water, sugar, etc.

(f) Destruction of mails by burning, acids, etc.

(g) Shooting of sentries.

(h) Stampeding of horses.

(i) Use of time bombs in cars, trains, etc.

ETC. ETC.

9. Sabotage, to be effective, requires the same degree of careful preparation as does military action. The first point is to choose an objective which has some value, even if it is only the sniping of a sentry or the firing of a stack of forage. Such shootings mean that the enemy must double his sentries or risk their loss; such destruction means more guards. So more troops have to be used, and this is one of your objectives.

The next step must be to study the place and conditions, so that the most favourable moment for success can be selected. A sure line of retreat, or an alibi, must be arranged beforehand. Often it will be necessary to wait a fortnight or longer before the right opportunity presents itself. At the same time, however, it may be necessary at times to carry out sabotage on the spur of the moment without previous preparation, for example when a convoy of lorries arrives unexpectedly in a village, and there is a chance of setting one on fire. Such opportunities should not be missed.

It is certain that the enemy will force a proportion of the inhabitants to work for him in mending roads, loading and unloading trains, and other works of a military nature. Such working parties provide good opportunities for sabotage by time bombs, by acids and other devices.

10. Organization :—

> (This particular pamphlet is intended simply for the use and instruction of guerilla 'parties'. The higher organization of guerilla warfare throughout a whole country or region is dealt with in the manual "*The Art of Guerilla Warfare*").

In the early stages of guerilla activities, before hostile counter-measures have become intense, it will be possible for the members of a party to live independently in their own villages and homes and carry on their normal occupations, only collecting when some operation is to be undertaken. The longer they can go on living in this way the better.

When the enemy begin to take active measures to prevent guerilla warfare by raids on suspected houses, by arresting suspects, etc., it will eventually be necessary for the guerillas to "go on the run" —i.e., to leave their houses and live out in the country, hiding themselves by day, and moving at night.

The number of men "on the run" in any one party must depend on the nature of the country. If it is wild, hilly, and forested, it may be possible for parties of up to 100 strong to avoid detection for long periods. If the country is flat and featureless and cultivated, it may be difficult for even one man to remain undetected for long. The organisation must therefore depend on the country ; the wilder it is the closer can the organization be -- i.e., the leader has his men closely under control all the time, and the party moves from place to place as necessary, to carry out operations or avoid capture. In less favourable country, the organisation must be looser, and men must be collected for action by secret means. If and when the enemy's activities make it too dangerous, for the time being, to continue, the men should leave their area, and join parties operating in more favourable conditions. These latter parties must always serve as a rallying point for men who have been forced by danger of arrest, to "go on the run" for deserters from the enemy, and escaped prisoners.

The leader is responsible for the organization; the importance of selecting only men who are reliable and resourceful is thus paramount.

11. Information:— If you can keep yourself fully informed of the enemy's movements and intentions in your area, you are then best prepared against surprise, and at the same time have the best chance for your plans to succeed.

G . S . (R)

The enemy is handicapped in that his men must wear uniform and are living in a hostile country, whereas your agents wear ordinary clothes and belong to the people and can move freely among them. Therefore, make every use of your advantage in order to obtain information. Suitable people must be selected from among the inhabitants to collect information and pass it on; these should be people who are unfit for more active work, but whose occupations or intelligence make them specially suitable for the task. The following are types who can usefully be employed:—

(a) Priests.

(b) Innkeepers.

(c) Waitresses, barmaids, and all cafe attendants.

(d) Domestic servants in houses where officers or men are billeted. These are a very useful source.

(e) Doctors, dentists, hospital staffs.

(f) Shopkeepers, hawkers.

(g) Camp followers.

These people must be trained to know what sort of information is required ; this is most easily done by questioning them on further points whenever they report anything, as they will then learn to look for the details required (see example at the end of the book). They must also be trained to be on the look-out for enemy agents disguised as compatriots.

It is important that as little as possible of this information should be in writing, or, if it is in writing, that it should not be kept any longer than necessary. All papers, documents, etc., dealing with intelligence or your organization in any way, must be destroyed immediately you have finished with them, or kept in a safe place until destroyed.

It has been proved over and over again in guerilla warfare that it is the capture of guerilla documents that has helped the enemy most in his counter-measures, These have been captured either on the persons of guerillas, or seized in houses that have been raided. The utmost care is therefore necessary.

12. Informers:— The most stringent and ruthless measures must at all times be used against informers; immediately on proof of guilt they must be killed, and, if possible, a note pinned on the body stating that the man was an informer. This is the best preventive of such crimes against the homeland. If

it is widely known that all informers will be destroyed, even the worst traitors will hesitate to sink to this depth of perfidy, whatever the reward offered.

If a person is suspected of being an informer, he can be tested by giving him false information, and then seeing if the enemy acts on it. If the enemy so acts, such evidence is sufficient proof of guilt, and the traitor must be liquidated at the first opportunity.

13. Enemy Counter-Measures and their frustration :— The best means of defeating the enemy's counter-measures is by superior information which will give warning of his intentions — i.e., of raids against suspected houses, of traps he may lay, of regulations he proposes to enforce in the territory he occupies, etc.

Attempts to bribe the people must be met by the measures shown in paragraph 12 above.

Certain counter-measures, however, can only be met by special action; for instance, the use of identity cards, which the enemy is certain to introduce when guerilla warfare becomes active, in order to assist him in tracing the guerillas. It will then be necessary to obtain or copy the official seals and stamps so as to provide identity cards for the guerillas.

When the enemy finds that passive means are insufficient to defeat guerilla operations, he will resort to active measures. These will probably take the form of mobile columns of considerable strength, horsed or in motors, including armoured cars and tanks, with which he will make sudden sweeps, often by night, through the various parts of the country. The bigger the column, the easier it is to obtain information about its projected movements, and it may even prove possible to combine several parties together and destroy it. If, however, the enemy's measures are so comprehensive as to lead to unnecessary risk, it will often be better for the guerillas to lie quiet for a month or so, or move to another district.

14. Conclusion:— All guerilla warfare and sabotage must be directed towards lightning strikes against the enemy simultaneously in widely distant areas, so as to compel him to weaken his main forces by detaching additional troops to guard against them. These strikes will frequently be most effective when directed against his communications, thus holding up supplies and eventually preventing him from undertaking large scale operations. At the same time, however, action should be taken against detachments, patrols, sentries, military lorries, etc., in such a way that the whole country is made unsafe except for large columns and convoys. This will hamper the enemy's plans effectively.

The civil population must be made to help by refusing to co-operate with the enemy, by providing information about the enemy, and by furnishing supplies and money to the guerillas. If they suffer inconvenience from your activities either directly or as a result of enemy counter measures, it must be explained to them that they are helping to defeat the enemy as much as their army at the front. The bolder the activities of the guerillas, and the greater the impunity with which they can act, owing to their careful planning and superior information, the more will the population despise the enemy, be convinced of his ultimate defeat, and help the guerillas.

Remember that you are fighting for your homeland, your mother, wife and children. Everything you can do to hamper and embarrass the enemy makes easier the task of your brothers-in-arms at the front who are fighting for you. As your activities develop, the enemy will become more and more ruthless in his attempts to stop you : the only effective reply to this is greater ruthlessness, greater courage, and an even, wider development of your operations. Your slogan must be "Shoot, burn and destroy". Remember that guerilla warfare is what a regular army has always most to dread. When this warfare is conducted by leaders of determination and courage, an effective campaign by your enemies becomes almost impossible.

THE PARTISAN LEADER'S HANDBOOK

G . S . (R)

ROAD AMBUSH. Appendix I

1. Planning.

   (a) Find out by what roads small detachments and patrols of the enemy are accustomed to move. Select on one of these roads a locality which offers a good opportunity for ambushing.

2. Locality.

The following points should be looked for in selecting the locality for the ambush:—

   (a) A line of retreat must be available which will give all the men a safe and sure way of escape. A thick wood, broken and rocky country, etc., give the best cover.

   (b) Firing positions are required which enable fire to be opened at point-blank range. When there is no chance of prior discovery by the enemy, it may some times be of advantage to improve the position by building a stone or sandbag parapet. This should not be done, however, unless it can be concealed from aircraft.

   (c) The locality should provide at least two fire positions and it is often better if these are on opposite sides of the road.

   (d) It is best if the fire position enables the approaching enemy to be in view for three or four hundred yards. By this means it can be discovered in time if the enemy is in greater strength than expected; in such a case the enemy should be allowed to pass without being attacked.

3. Information.

Then get the following information :—

   (a) Do the detachments move on foot, mounted, or in motor vehicles?

   (b) What is the average strength of these detachments? How are they armed? How many vehicles?

   (c) Do they use armoured cars and light tanks to patrol the roads?

   (d) At what times do they pass the place you have chosen ?

- (e) Do they move in one block, or do they put men out in front and behind to guard against surprise? How do these men move, and how far from the main body?

- (f) How will they try to summon assistance if attacked? Where is the nearest place such assistance can come from?

- (g) If the detachment is carrying supplies, are those supplies of a type which can be easily destroyed by you, or be of use to you?

- (h) What sort of troops are they, active or reserve, elderly, young, or what? Is there an officer with them? Can he be picked out and shot by the first volley? Can the N.C.Os be picked out as well?

4. Action.

- (a) The men must get into position without any chance of discovery. If there, is any doubt, the position should be occupied by night,

- (b) Sentries must be posted to give warning of the enemy's approach. They must be in sight of the firing position. It is not necessary to use guerillas for all sentry posts; a woman or child can sometimes be employed with advantage as they need not be in hiding.

- (c) A simple system of signalling by sentries must be arranged. This can be the removal of a hat, doing up a shoelace or any natural action of that nature.

- (d) If the enemy detachment is preceded by scouts, or a scouting vehicle, these should be allowed to pass on and not be fired at. Sometimes, however, it may be advantageous to place one or two guerillas further on from the firing position to shoot these scouts. They must never be fired on, however, before the main attack begins; the guerilla leader must make certain this is known and understood.

- (e) The leader must give the signal to open fire. This can either be pre-arranged or given at the moment. Fire must be rapid fire, so as to have an immediate overwhelming effect.

- (f) Two or three of the best shots must be detailed to shoot any officers or N.C.Os. If these cannot be recognised by their uniform, they can be discovered by noting who is shouting orders, etc.

G.S.(R)

(g) If the enemy appears to be destroyed, and it is intended to destroy or loot any cars or lorries, men for this task must be detailed beforehand. The rest must remain ready to open fire in case enemy are concealed in the lorries, or reinforcements arrive.

(h) The leader must give the signal to retire, and this signal must be unmistakable. To judge the correct moment to break off the action is the leader's most difficult task. If the opening volleys of fire have not disorganized the enemy, it will probably be better to retire immediately, and be content with the damage done. If, however, the enemy detachment is completely destroyed, the opportunity should always be taken to seize all rifles, ammunitions, etc., and destroy or loot all other material. All papers and documents found should be taken away for examination. The dead must be searched for anything that may be useful.

(i) Remember that soldiers will always face the direction from which they are being fired at. It is usually best therefore to divide the party into two groups, on different sides of the road, of which only one group should fire first. The enemy will then, face towards this group and start to attack and fire. The other group must then shoot the enemy in the back.

(j) Sentries must remain in position until the leader gives the signal to retire.

(k) Retirement when begun should be as rapid and dispersed as possible, i.e., the party must break up, and collect again as the leader may have ordered. Make full use of the time until the enemy hears of the attack to get right away from the scene.

(l) All wounded guerillas must he carried away if possible. It may be useful to have a few horses hidden at a short distance to carry wounded.

5. Road Blocks. The use of road blocks by means of trenches, felled trees, rocks, etc., in conjunction with an ambush must be carefully considered.

At the commencement of guerilla warfare, before the enemy has had experience, it may be useful to have a block at the place of ambush, so as to force lorries to halt. When, however, the enemy is experienced, he will use scouts and patrols on all roads, and these will be warned by the blocks and so warn their detachments. A stout wire rope fastened across the road after scouts have passed, at a suitable height to catch the motor driver, is a useful device.

If it can be arranged to have mines or bombs buried in the road, which scouts will not, see, these are of great .assistance in demoralising the enemy. Fire should not be opened until the mine has been exploded under the enemy. Here are details of road mines :—

(1) Crater or Land Mine :-- 60lbs of high explosive buried 5 feet deep and fired electrically will produce a crater 25 feet across which will wreck a tank, armoured car, etc. completely. All traces of the digging must be obliterated carefully to avoid the enemy locating the ambush by scouts on motor cycles. The digging for this reason should be done outside the tarred area of the road but close to it.

If done overnight and watered as soon as filled in, the traces of excavation can best be obliterated. The debris from the explosion will be thrown as far as 200 yards. Men in ambush 100 yards away, behind cover, are sufficiently protected, however. A crater so formed, if in a defile, is an impassable obstacle to tanks. The method of laying the charge is as follows :—

A hole 5 feet deep is dug and 60 1bs. of explosive in its paper wrapping is placed in the bottom. The paper wrapping of one packet is broken and an electric detonator is inserted, dug well into the explosive itself. Two wires, 100 yards in length, are joined to the two ends of wire projecting from the electric detonator.

If the ground is wet, these joints must be protected with insulating tape or some other covering. Care must be taken that the two joints do not touch. The long wires are then led away to a distance and hidden where necessary in a shallow trench. When the two outer ends of the wire are connected to the terminals of an ordinary car battery the charge will be exploded.

(2) Small Land Mine;— A charge of 10 lbs of high explosive with not more than 6 inches of earth covering will blow the track off a tank or the wheel off a lorry passing over it. The road should therefore be partially blocked by a broken-down farm cart or other means so that all traffic is forced to proceed through a very

G.S.(R)

limited gap. The charge should be placed as in the preceding paragraph, but may be fired from a distance of 25 yards. Care must be taken to judge the exact moment the wheel of the vehicle passes over the charge.

(3) Hand Bombs:— These are of two sorts, those with a 7 second time fuze and those which go off on impact. The impact bomb is essentially for throwing. The thrower should locate himself behind a wall or other cover, preferably within 10 yards of where the enemy vehicle will pass. The bomb will smash through any metal it is in contact with on impact, but will have little effect on a vehicle elsewhere. It should therefore be thrown to hit the side or tracks of a tank or the wheel of a vehicle. The time bomb is most effectively used for the destruction of any machinery or vehicles in which it is placed or thrown.

6. Remember :—

(a) Let scouts pass.

(b) Use your best shots to kill any officers or N.C.O.s and drivers of vehicles immediately.

(c) Armed sentries must remain at their posts until ordered to retire.

(d) Any looting or destruction must be protected by men ready to fire.

THE PARTISAN LEADER'S HANDBOOK

G . S . (R)

RAIL AMBUSH   Appendix II.

In general the rules for road ambushes apply to rail ambushes, so read them and make certain you understand them.

The difference between a rail ambush and a road ambush is that in a rail ambush you must combine some plan to wreck the train, either by derailng it, by blowing a mine under the engine, or other means. It is not sufficient merely to shoot at the train ; this would do more harm than good and must be avoided.

(1) The principle is first to derail the train and then shoot down the survivors.

(2) Choose some place which is suitable for wrecking, for example a high embankment where the falling engine will drag the coaches down with it: or a bridge, where the train will, with luck, fall into the river.

(3) Do not choose a place where trains run slowly ; the faster the train is going, the better results you will get.

(4) The coaches at the rear of the train will probably suffer least damage; your first volleys should be directed against them.

(5) It is best to dispose your party in two groups, as in a road ambush, on opposite sides of the train.

(6) The signal to shoot will be when the wrecking starts or the mine is exploded. Everyone must start firing immediately.

(7) The train must not be looted until you are certain that all resistance by the enemy is at an end. After looting, it should be set on fire.

(8) If the train is armoured, and the wrecking has not been severe, it may be better to retire immediately. An armoured train will usually have many machine-guns with it.

Read again the rules for a Road Ambush and apply them to this case.

Here are some methods of derailing a train :— To derail a train with certainty, both rails must be cut. This can be done very easily in the following ways:—

(1) One pound of high explosive pressed hard against the side of each rail.

(2) Three pounds of high explosive placed against the underside of each rail.

(3) Ten pounds of high explosive buried under the ballast not more than 4 inches from each rail.

(4) A single charge of fifty pounds of high explosive buried three or four feet deep between the rails. This will lift the locomotive ten feet into the air and is the best way where no bridge or steep slope can be found.

If the derailing is done by methods (3) or (4), or where the ballast has been allowed to come close up under the rails by method (2) as well, it will be possible to lay the charge so that it will be undetected by day. Care must be taken not to show any signs of digging. A tin of water should be carried to wash down the stone ballast and clean it of earth adhering to it when using methods (3) or (4).

In all cases, it is best to fix the charge under or just in front of the front wheels of the locomotive. This can be done in two ways :—

(a) By means of an electric detonator with long wires leading to a battery where a man is concealed to operate it at the right moment.

(b) By means of a striker machine which is buried under a sleeper next to a rail joint. The weight of the locomotive passing over releases a striker which fires the charge by means of an instantaneous fuze.

In both cases, the detonator must be buried firmly in the explosive. When a battery is used, great care must be taken that the battery does not come near the ends of the wire till the last moment, to avoid accidents.

A length of wire up to 100 yards may be used leading away from the explosive to a hidden spot where it is fired. Insulated wire such as is used for electric light in houses must be used. The accumulator battery out of a car is best but a good hand torch dry battery will do.

G.S.(R)

Diagrams of methods (1) and (2) of cutting rails:—

Method (1).

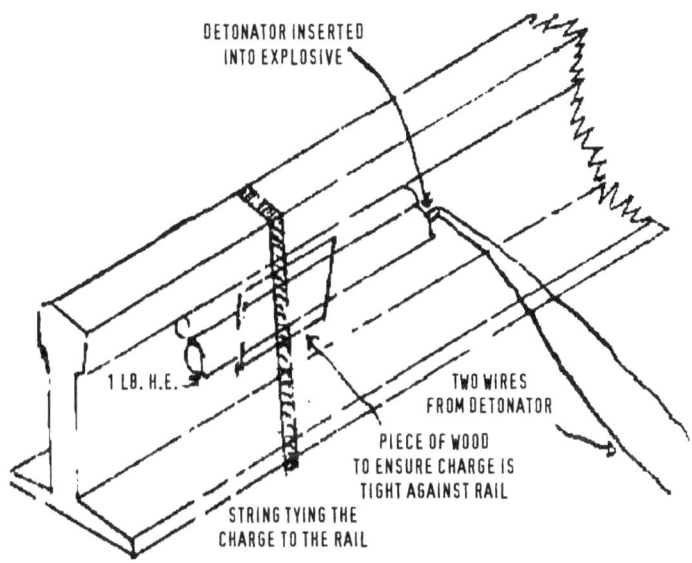

Method (2).

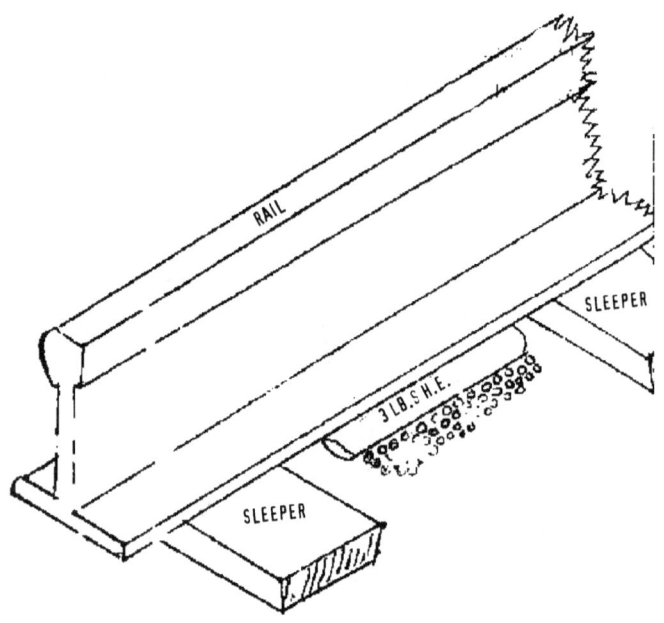

Detonator and wires to battery should be arranged as in Method (1).

G . S . (R)

Method of connecting electric detonators.

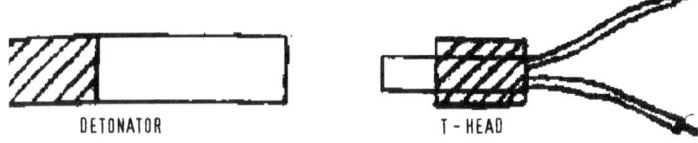

DETONATOR   T - HEAD

The electric detonator is in two parts :--

(1)   The detonator, which is a small copper tube closed at one end and open to the other.

(2)   The T head, which has two wires sticking out of one end and a very thin bridge of wire like the filament of a lamp the other. The filament end of the T head is pressed into the open end of the detonator. When an electric current passes through the filament it gets red hot and burns away completely, but in doing so, ignites the detonator.

When wires are joined together or to the T head or battery, the covering must be cut away and the metal cleaned bright by scraping with a knife. The wires may then be twisted together. The bare wire at a joint must never touch anything, especially another joint. It is best to bind with insulating tape or a piece of cloth. The joining of the wires for two charges fired by one battery is shown later.

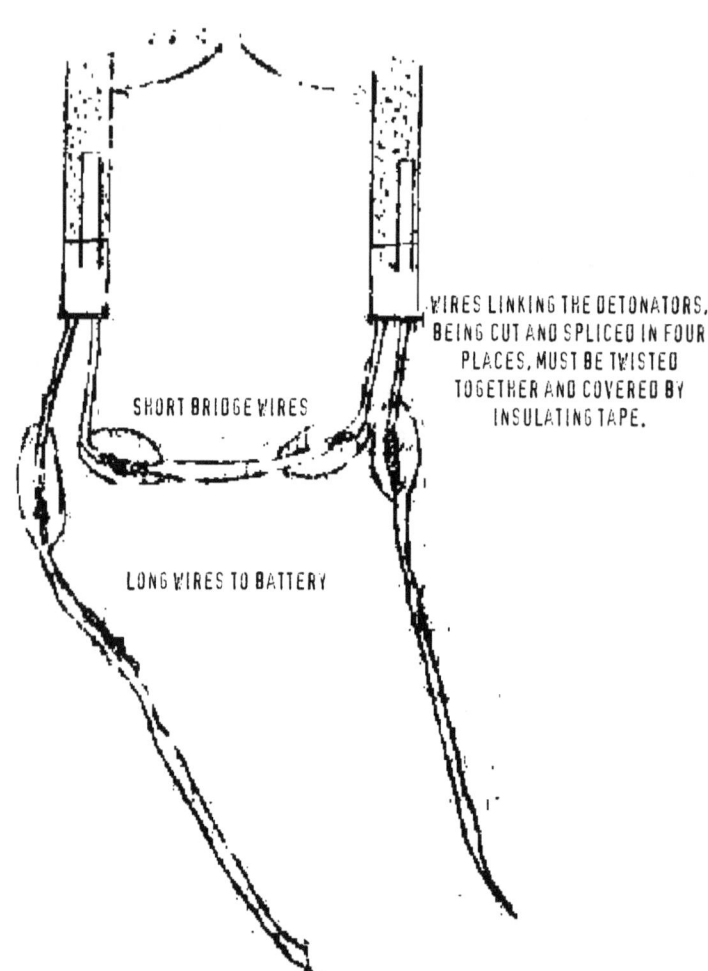

G. S. (R)

Destruction of Railway Engines:—

(1) If you have no explosives, run off most of the water in the boiler and bank up the fire. The fire box, no longer cooled by the water, will get red hot and the steam pressure will bend it in.

(2) If you have explosives, make it up into one pound packets each with a hand bomb time striker mechanism. This mechanism will explode the charge six seconds after the pin is pulled out. The best places to put these charges are on any of the large machined portions of the engine which the hand bombs will cover and are not more than 1" thick. If the engine is cold, open the smoke box in front and put a charge just inside one of the tube openings.

THE PARTISAN LEADER'S HANDBOOK

G.S.(R)

THE DESTRUCTION OF AN Appendix III.
ENEMY POST, DETACHMENT
OR GUARD.

1. The object of this can be either to inflict causalities on the enemy, or to carry out the destruction of some place which the detachment is guarding.

2. The detachment will usually be housed in a small house, hut, or tents, and will have taken steps to try and make these safe against attack. Remember, however, that if you use cunning, patience, and determination, no small post can be made impregnable and at the same time do its job of protection properly.

3. Information :-- You must get detailed information of the posts in your area, and then decide which offers the best chance of success. It may not be possible to get full details of all, but you will get enough information about some of them to enable you to select one and carry out a successful attack.

4. The points on which you should get information are :--

   (a) Strength of the detachment, number of officers, N.C.Os etc.

   (b) Who commands the detachment?

   (c) Are the troops active or reserve? Are they old or young men? To what regiment or district do they belong?

   (d) What arms and equipment do they carry? Have they machine-guns?

   (e) Is there a reserve of arms in the post? Where are they kept?

   (f) What are the orders for safe custody of arms? Are any locked up?

   (g) What means of communication has the post got — i.e.
       (i) Telegraph or telephone or wireless.
       (ii) Signal flags.
       (iii) Rockets.
       (iv) Pigeons.
       (v) Sirens, hooters.
       (vi) Messengers. ETC.

       Can any of these be destroyed when necessary?

(h) What sentries does the post provide —
 (i) On the rail way, bridge, or store it is guarding?
 (ii) On the post itself?

(j) At what hours are sentries relieved —
 (i) By day?
 (ii) By night?

(k) How is relief carried out?

(l) Is there a group of men in the post always ready for immediate action? How strong is it?

(m) How long is each sentry's beat? What are its limits?

(n) What places can these sentries not see except by going to them?

(o) Are any civilians allowed to approach or enter the post, selling food, papers. etc.? Can you use any of these civilians to get information?

(p) Are there any searchlights in position?

(q) Is the post protected with barbed wire? Is this wire electrified? How do soldiers get in and out?

(r) Where does the post get its water supply from? Can the source of water be destroyed?

(s) How often is the post and its guard inspected by someone from outside?

(t) How far away is the nearest reinforcement, and how long would it take to come? Can it be ambushed on the way by another party?

(u) Can your destructive work be undertaken while the post is being fired at, or must the post first be destroyed completely?

(v) Can the post be blinded by smoke bombs for long enough to allow the destruction to be done?

G . S . (R)

(w) Are there watch-dogs, alarm traps, etc.?

ETC.   ETC.

5. Plan :— This must depend on the information collected regarding the daily life and habits of the post, the state of alertness of the guard, its strength, armament, etc.

If the post is very small — say six to eight men — it may be possible to capture it by getting one or two men inside to seize the arms and hold up the guard at the moment the sentries are shot; on the other hand, it may be possible to rush the post from outside after shooting the sentries, to surround it and cut all communication, and shoot down all the men inside. It will also frequently be practicable to carry out destruction by one group while the other group of the party prevents the enemy of the post interfering. This depends to some extent on how long the destruction will take.

If the post is large, it will probably not be possible to destroy it. In such cases, if you wish to carry out some really important destructive work, it should be attempted by masking the post with heavy fire, smoke, etc. Such an attack usually has most chance of success when carried out by night.

> In every case of an attack on a post, your first care must be to arrange for the destruction of means of communication — i.e., telegraph, wireless, etc.—unless you have a plan to ambush reinforcements.

Do not alarm any post that you mean eventually to attack —i.e., do not allow men to snipe it to cut off its water supply, etc. Leave it absolutely quiet until the moment for attack comes. This will put the enemy off his guard.

THE PARTISAN LEADER'S HANDBOOK

G.S.(R)

CONCEALMENT AND CARE OF   Appendix IV.
ARMS AND EXPLOSIVES.

Try and get your arms before the enemy invades your country, so that you can conceal them more easily and at leisure.

1. One of the first acts of the enemy will be to demand the surrender of all arms held by the civil population.

2. All arms, bombs, etc., which are concealed must be protected against damp, rust, etc. ; remember that your life and that of your friends may depend on a weapon in good order. The best way of preserving rifles, revolvers, etc., is to cover them with mineral jelly or vaseline, and wrap them in greasy paper or cloth. They may then be safely buried.

3. Places where arms can be concealed are:—
   (a) In the ground by burying. Choose a place where the earth has already been turned, or else go far away into a wood, etc.
   (b) In the thatch or roof of a house.
   (c) In a well-shaft, by making a chamber in the wall six feet or more down the shaft.
   (d) In the banks of streams, in hollow trees, behind a waterfall, etc.
   (e) In haystacks, potato or turnip heaps, ditches, culverts, etc.
   (f) Do not use places like cellars, wooden floors, cattle sheds, etc., which the enemy is. bound to search.
   (g) As a last resort, give them to your women if caught unexpectedly.

4. You must make every effort to obtain arms and ammunition from the enemy during ambushes, raids, sniping, etc., as it will be difficult in time of war to replenish your stock by other means. Boxes of files and ammunition are frequently transported by rail and in lorries, inadequately guarded; find out when these are being carried and try And get them. Be very cautious of buying arms from a supposed enemy traitor. This is a common way of inducing you to walk into a trap.

THE PARTISAN LEADER'S HANDBOOK

G.S.(R)

THE ENEMY'S INFORMATION SYSTEM AND HOW TO COUNTER IT.   Appendix V.

As soon as guerilla warfare or sabotage commences the enemy will set up an information organization in order to try and find out your organisation, leaders and intentions.

The methods he will employ are as follows :—

(1) Local agents, selected from amongst the inhabitants, are either bribed or compelled to act for him,

(2) Agents recruited from his own or other countries and imported into your area. These two types of agents can only be discovered by very careful work on your part, by getting information regarding arrivals of unknown people, by laying traps for suspected agents, etc.

(3) Special information branches that he will form.

(4) Captured prisoners and their interrogation.

(5) Captured documents which may reveal details of your organization, plans, names of partisans, etc. It is most important that no documents should be kept unless absolutely essential, and these should never be carried on the person for longer than necessary. This is usually the enemy's best source of information.

(6) Censorship of civilian letters.

(7) By placing agents among captured partisans. This is a difficult thing to counter and can only be met by strict discipline among the partisans in the prisons and concentration camps. They should be trained never to talk about their military matters, to mention names, or to give away any information at all. Steps must be taken within prisons by the partisans to test and try out every prisoner who comes in, to make absolutely certain that he is not an agent in disguise.

(8) Listening sets:— These will also be placed in prisons and camps, so all conversation must be restricted to general matters and nothing said which might lead to the capture or death of your compatriots.

(9) Men who are captured must at once organize themselves in the prison to censor all their own letters that they are writing, to friends outside, and to censor all incoming letters to individual prisoners.

(10) The best method of dealing with informers is their ruthless extermination when discovered, as described in the main part of this book.

(11) Prisoners who are being interrogated may be tempted, by the fact that there is only one enemy in the room, to give away information if pressed, as they may feel that only one person will know it. All men must know that this is not correct ; not only will the enemy install listening sets in the room in which the prisoners are interrogated in order that two or three people may hear any confession, but also all the information a prisoner gives, and his name and district, will he taken down in writing and distributed everywhere. His comrades would then eventually discover his treachery and he would he dealt with suitably when the enemy has been defeated.

You must try and break up or hinder the enemy's information organization by all means. The most effective is the destruction of the personnel engaged on that work. Intelligence officers, N.C.Os, etc., will frequently work individually and move about the countryside. Opportunities must be sought to kill them and destroy or carry off any papers they are carrying.

G.S.(R)

HOW TO COUNTER ENEMY ACTION. Appendix   VI.

The enemy will make use of his superior armament to try and break up guerilla, activities. Here are some of the methods he will employ, and ways for you to counter them :—

(1) Aeroplanes:— These will be used to search the country for guerilla parties, and possibly also to attack them. The best counter is concealment, therefore, move as much as you can by night. By day on the approach of an aeroplane, men must be taught to get under whatever cover is available, and to lie still with faces to the ground. Movement and human faces show up to aeroplanes at once.
Do not fire at an aeroplane unless actually attacked by it. Remember that an aeroplane, if it sees you, will at once report your position to the nearest military detachment who will come out after you. Therefore, if you think your party has been seen, move off at once to some other place, and keep a good look-out.

(2) Tanks, armoured cars, armour-plated lorries etc. :—   Do not shoot at these haphazardly, it will have, no effect unless you have anti-tank rifles and bombs, etc. You must lay a proper trap if you are trying to destroy them — i.e. a road mine or block, or the vehicle must be halted. Remember that these vehicles shut down their windows when attacked, and are then very blind ; it will then be possible for bold men to crawl close enough to bomb them or set them on fire with petrol.

(3) Gas:— The enemy will only use gas if he gets you in a corner and other methods fail. Therefore your first precaution must be to avoid being caught where you cannot get away. Your information of the enemy's plans and proper posting of sentries and look-outs when the party is collected will prevent you being caught. If you hear that the enemy intends using gas against guerillas, all men should provide themselves with gas-masks.

(4) Shells, bombs, grenades:— Against these weapons the best protection is to be down flat behind any cover available, such as a bank, ditch, etc,

(5) Machine guns, etc.:— Smoke bombs can be used to create a smoke screen between yourself and the machine-gun so as to enable you to get away.

G.S.(R)

GUERILLA INFORMATION SERVICE. Appendix VII.

1. Early information of the enemy's moves, strength, intention, etc. is vitally important. You must therefore impress on all your compatriots the necessity of passing on to some members of the party any information they hear. The following, owing to their occupations, are in a good position to get news :—

(a) Innkeepers, hawkers.

(b) Waitresses, barmaids, etc.

(c) Postmen, telephone and telegraph operators, (d) Station-masters, railway porters and staffs.

(e) Doctors, priests, dentists, hospital staff.

(f) Domestic servants, barbers.

(g) Shopkeepers, newsagents.

(h) Contractors, camp followers, camp sanitary men.

(i) All people who have access to military camps, establishments, etc.

(j) Discontented enemy soldiers.

2. Domestic servants and cafe attendants are particularly valuable agents ; they must be encouraged to gain the confidence of the enemy soldiers, and be on easy and intimate terms with them. Suitable agents of this type should be introduced into houses where enemy officers are billeted, etc. It is a natural weakness of soldiers in a hostile country to react favourably to acts of courtesy and kindness from women; such men will frequently drop unsuspecting hints that they are shortly going on patrol, etc. The agent must then find out as much detail as possible and pass it on at once.

3. Discontented soldiers must be discovered, i.e., those who have recently been punished, have had their pay stopped, etc. These, if encouraged, may give useful information.

4. Information should be passed by word of mouth unless that is impossible. If impossible, it must be written and sent by messenger (children frequently make good messengers) or placed in a pre-arranged place, and then destroyed by the recipient.

G. S. (R)

SABOTAGE METHODS. Appendix-VIII.

Sabotage means any act done by individuals that interferes with the enemy and so helps your people to defeat him. It covers anything from the shooting of a sentry to the blowing-up of an ammunition dump. The following are various acts, and the best way of carrying out the difficult ones :—

(1) Lorries, cars, tanks, etc. :-- Burn them by knocking a hole in the bottom of the petrol tank, and setting fire to the escaping petrol. If you can't burn them, put water or sugar in the petrol tank, or remove the magneto, etc. This will temporarily disable the vehicle.

(2) Munition Dumps:— The best method is to lay a charge of explosive among the shells and then explode it, but it will be rare that you will get an opportunity to do this unless you are disguised as an enemy soldier. There are other ways. If the dump is in a building, a good way is to set fire to the building. Use oil-soaked rags, shavings, thermite bomb.

If the dump is in an open field or by the road, throw a special bomb into it (this must be a bomb with at least one kilogramme of explosive in it, and you must hit a shell or it will not be effective).

(3) Cement :— Open the sacks, and pour water on them, or leave them for rain and moisture to get in.

(4) Hay, Forage:-- Burn or throw acid or disinfectant.

(5) Petrol stocks:-- Use a special bomb or thermite bomb.

(6) Refrigerator sheds, and refrigerator railway vans:— Destroy the refrigerating apparatus.

(7) Sniping and killing sentries, stragglers, etc. Get a rifle or revolver with a silencer, but use a knife or noose when you can. This has a great frightening effect. Don't act unless you are certain you can get away safely. Night time is best and has the best effect on enemy nerves. Get used to moving about in the dark yourself. Wear rubber shoes and darken your face.

(8) Telegraph Lines on roads and railways:—

Cut these whenever possible. When you cannot reach them, throw over a rope with a weight on the end and try and drag them down. Cut down a tree so that it will fall across them.

(9) Railways:— Jam the points by hammering a wooden wedge into them. Cut signal wires. Set fire to any coaches and wagons you can get at. If you can use explosive, try and destroy the points. Remember that railways can carry very little traffic if the signalling apparatus is interfered with, and this traffic must go very slowly.

(10) Water Supplies:— Contaminate water which is used by the enemy, use paraffin, strong disinfectants, salt, etc,

(11) Destruction of leading marks, buoys, lightships, etc. in navigable waters.

(12) Burning of soldiers' cinemas, theatres:—
Cinema films are highly inflammable. The cinema should be fired during a performance by firing of the films in the operator's box. This should easily be arranged.

(13) Time bombs, cigar-shaped, are very suitable for placing in trains, lorries, etc. They are made of lead tubing, divided into two halves by a copper disc. Suitable acids are put in each half, and when they have eaten the copper away, the acids combine and form an intensely hot flame, which will set fire to anything with which it comes into contact. The thickness of the copper disc determines when the bomb will go off. Get some of these bombs.

FINIS

www.ingramcontent.com/pod-product-compliance
Lightning Source LLC
Chambersburg PA
CBHW050246230526
45470CB00005B/2129